Moldy

The Protists & Fungi Files

Discovery Channel School
Science Collections

© 2001 by Discovery Communications, Inc. All rights reserved under International and Pan-American Copyright Conventions.
No part of this book may be reproduced in any form or by any electronic or mechanical means, including
information storage devices or systems, without prior written permission from the publisher.
For information regarding permission, write to Discovery Channel School, 7700 Wisconsin Avenue, Bethesda, MD 20814.
Printed in the USA ISBN: 1-58738-133-8

1 2 3 4 5 6 7 8 9 10 PO 06 05 04 03 02 01

Discovery Communications, Inc., produces high-quality television programming,
interactive media, books, films, and consumer products. Discovery Networks, a division of Discovery
Communications, Inc., operates and manages Discovery Channel, TLC, Animal Planet, Discovery Health Channel, and Travel Channel.

Writers: Jackie Ball, Stephen Currie, Julie Danneberg, Kathleen Feeley, Katie King, Lisa Krause, Uechi Ng, Monique Peterson, Darcy Sharon, Denise Vega.
Editor: Katie King. **Photographs:** moldy bread, PhotoDisc; p. 2, sulfur shelf, Corel, moldy bread, PhotoDisc; p. 5, kelp, Corel, green algae, ©Cabisco/Visuals Unlimited, euglena, ©David M. Phillip/Visuals Unlimited, mushroom, Corel, sulfur shelf, Corel, moldy bread, PhotoDisc; pp. 6–7, egg slime mold, Corel; p. 8, waxy cap, sulfur shelf, egg slime, turkey tail (all), Corel; p. 9, blue fungi, winter polypore (both), Corel; p. 11, Dr. Carl Woese, courtesy of Carl Woese, deep sea vent, ©WHOI, J. Edmond/Visuals Unlimited; p. 12, amoeba, ©M. Abbey/Visuals Unlimited, paramecium, ©Karl Aufderheide/Visuals Unlimited; p. 13, euglena, ©David M. Phillips/Visuals Unlimited, *Vorticella*, © Michael Abbey/Visuals Unlimited, algae, Cabisco/Visuals Unlimited; pp. 16–17, potato famine engraving, Brown Brothers, Ltd.; pp. 18–19, pond water, Corel; p. 19, paramecium, ©M. Abbey/Photo Researchers, Inc.; p. 20, sushi, PhotoDisc, bread, Corel, ice cream, PhotoDisc; 21, cheese, Corel; 22, deathcap, Corel, Caesar's mushroom, ©John Serrao/Photo Researchers, Inc., destroying angel, Richard Thom/Visuals Unlimited; p. 23, morels, Corel, truffle, Rob & Ann Simpson/Visuals Unlimited, pig, Corel; p. 26, Alexander Fleming, Brown Brothers, Ltd.; p. 27, penicillin in petri dish, Larry Jensen/Visuals Unlimited, drugstore sign, ©UPI/Corbis Bettmann; p. 28, otter, diver (both), Corel; p. 29, kelp, sea lion, puffballs (all), Corel; p. 31, cheese, PhotoDisc, earth star, ©Jeff Lepore/Photo Researchers, Inc., fairy ring, Wally Eberhart/Visuals Unlimited, lichen, Corel. **Illustrations:** pp. 14–15, compost heap comic, Jack Desrocher. **Acknowledgments:** pp. 16–17, Irish potato famine, W. Steuart Trench and Catherine Hennagan excerpts reprinted from THE IRISH FAMINE by Peter Gray © 1995 Harry N. Abrams; p. 21, excerpt from MY SIDE OF THE MOUNTAIN by Jean Craighead George, © 1959, 1988, Dutton.

CONTENTS

Moldy

Two Great Kingdoms

Look very closely. Many living things on Earth are neither plants nor animals. Some of them are protists and fungi, two different kingdoms. Don't worry—these are *not* alien creatures!

What are they? One protist is seaweed, the plantlike algae you see at the beach. Others—like the protozoa you see in ponds—are animal-like. Slime molds that slither over forest logs are protists, too. Fungi include mushrooms and yeasts and some kinds of molds—like the kind you see on bread. But it doesn't stop there. Protists and fungi are two diverse kingdoms. Despite their diversity, protists and fungi eat, grow, and reproduce like all living things.

In Discovery Channel's MOLDY, you'll learn why plants and animals cannot survive on Earth without protists and fungi.

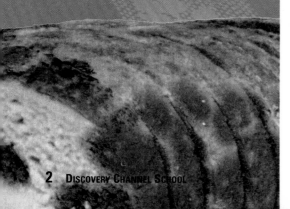

The Protists & Fungi Files

Protists & Fungi 4
At-A-Glance   Find out why protists and fungi live in their own special kingdoms.

Ready for Slime Time 6
Q & A   A slime mold talks about its colorful life.

It's All Relative 8
Almanac   Protists and fungi make up a large part of Earth, especially compared to plants and animals.

All in the Family Tree 10
Scientist's Notebook   See how Carl Woese has reorganized a traditional tree of life.

Keep on Moving 12
Picture This   Protozoa are protists that know how to get around. Find out how they move, eat, and survive.

Having a Breakdown 14
Timeline   Inside a compost pile fungi turn that trash into "treasure" for all of us.

Disaster in the Wind 16
Eyewitness Account   Read about how one fungus had a powerful effect on Ireland's history.

A Paramecium's Paradise ... 18
Virtual Voyage   You're a protist zipping around a pond and experiencing many interesting adventures.

Good For Us 20
Scrapbook   Believe it or not, protists and fungi are in the foods you eat every single day.

To Eat or Not to Eat 22
Amazing But True   Enter the world of mushrooms and other treats. Many are delicious—but a few are deadly.

The Protozoan Zone 24
Map   Find out how some protists take their toll on humans around the world.

The Miracle of Mold 26
Heroes   Sir Alexander Fleming discovered properties of a fungus that saved millions of lives.

The Mystery of the Failing Fishing Hole 28
Solve-it-Yourself Mystery   Marnie Fisher and her grandfather have gone fishing, but they can't catch a thing. See if you can find out why their favorite fishing hole is empty.

Slimy Stuff 30
Fun & Fantastic   Get a load of the largest fungus ever recorded and many other fun fungi and protist facts.

Where do fungi go for a good time? Find out on page 14.

Final Project
Classify Your World 32
Your World, Your Turn   Research scientific classification systems and debate which one is the best.

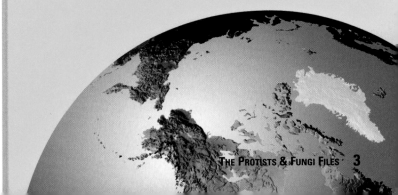

AT-A-GLANCE
Protists

Take a quick stroll along a typical city street. You might see fresh fruits and vegetables piled up in a grocery window; squirrels scampering on tree limbs; and kids and adults jogging, walking, or riding scooters. The longer you walk, in the city or the country, the more examples you'll see of the same two kinds of living things: animals and plants. Plants and animals. It's enough to make you think that everything alive must be one or the other.

But that's not the case. Not all living things fit perfectly into the animal or plant kingdoms. So many other life-forms exist that some scientists argue there should be a total of five kingdoms: plants, animals, fungi (FUN-jeye), protists, and monerans (bacteria).

Single-celled bacteria were the earliest forms of life, originating on Earth billions of years ago. Which life-forms came next? Some scientists believe that protists first evolved from bacteria. As a group, protists are the most diverse. Some, called protozoa, act and look like animals. Others act and look like plants. Some of these plantlike protists are unicellular and others are multicellular. If protists are so diverse, why are they all lumped together into one kingdom? One reason is simply that they don't fit in anywhere else.

Scientists place fungi in their own kingdom because fungi are a diverse group of organisms. Yeasts are minuscule and single-celled. But mushrooms—the fruiting bodies of some fungi—are multicellular. Some mushroom varieties can be gigantic—large enough to cover a field or encircle the thick trunk of a 200-year-old tree. Fungi absorb nutrients from living or, more often, dead organic matter.

Together protists and fungi play a key role in all ecosystems. Without them, few plants or animals could exist. Read on to learn more about these fascinating living things.

TAKE FIVE

Some scientists classify living things into five kingdoms.

- MONERANS (bacteria)
- PROTISTS
- FUNGI
- PLANTS
- ANIMALS

& Fungi

Protists

Macrocystis kelp

Derbesia green algae

Euglena

Fungi

Sulfur shelf (*Laetiporus sulphureus*)

Mold growing on bread

Marasmius scorodonius mushrooms

Ready for Slime Time

Yellow slime mold

A protist talks about life in a crazy mixed-up kingdom.

Q: You're a yellow slime mold. Love that color! You're a carpet of sunshine . . . a parade of tiny daffodils . . .
A: A blanket of buttercups, I know, I know. I've heard it all. Sometimes it's a burden to be so beautiful, but having slime in my name keeps me modest.

Q: Are other slime molds as attractive as you?
A: Each in its own way. Slime molds come in lots of different colors, from bright orange to electric blue. We have interesting shapes, too—twisty pretzels and fluffy scrambled eggs. Yes, we're a handsome group, and a hard-working one, too.

Q: Hard-working? You don't look very busy to me, draped over that log.
A: Appearances can be deceiving. I may look as if I'm relaxing, but I'm actually working very hard.

Q: Doing what?
A: Right now I'm reproducing. Before that I was eating. This decayed log is my food supply, and even as we speak, I'm breaking down food to make all the energy I need.

Q: But why do you need a log for lunch? I thought plants could make their own food.
A: They can. But I'm not a plant. I don't photosynthesize for food.

Q: So—you're not a plant. Are you an animal?
A: No. I move around to get food in the way animals do—in my case it's more like a slow creep—but I reproduce with spores like a fungus.

Q: So if you're not really an animal or a plant, who are you?
A: Think about it. Animals and plants aren't the only life-forms in this world.

Q: But what else *is* there?
A: There's bacteria. Scientists classify them in their own kingdom. And there's the fungi kingdom, which contains all different kinds of fungus:

mushrooms, spores, molds, yeasts. They're organisms that take their energy from other matter—sometimes living but mostly from dead things. Everything from the blue mold on that sandwich you left in your locker to the athlete's foot fungus snacking on the stuff between your toes.

Q: So you must be a fungus, right?
A: Close, but wrong again. I have fungus-like characteristics. I produce spores, for example, but I am not part of the fungi kingdom.

Q: I give up. Where do you fit in?
A: I'm a fungus-like member of the protist kingdom—and the beauty of my kingdom is that nothing really fits in! We're different from all the other life-forms, and that's a big reason certain scientists agree we should be grouped together.

Q: So why are you so special?
A: Well, protists like me don't make our own food, but we can always find plenty to eat. We break down matter to its simplest forms, which helps the earth reabsorb it easily.

Q: So how are you different from everyone else?
A: Some protists move and hunt for food like animals. They're called protozoa. You've probably never seen any, because they're so tiny. They're made of only one cell. But trust me—every drop of pond water contains lots of them.

Q: And what about the other protists?
A: Other protists act like plants. They have chlorophyll, and they photosynthesize. Take algae—and there's plenty to take! Seven thousand species of green algae alone, all busily cooking up dinner every time the Sun shines. Not to mention red algae, and brown algae-like kelp, which looks like seaweed. Don't forget diatoms and dinoflagellates (deye-noh-FLAJ-uh-lihtz). All algae species have chlorophyll.

Q: Are there any protists that act like both animals and plants?
A: Sure. Euglena, for one. It is an example of a one-celled alga that has characteristics of both plants and animals.

Q: So back to you again. You're a fungi-like protist?
A: Yup. As I said earlier, I feed on bacteria and decaying matter. I look like a fungus, but my cell structure is different. I'm part of the protist kingdom.

Q: So do you like what you do?
A: Of course. I do good work. I should—I've been at it a long time.

Q: How long?
A: Billions of years, ever since protists evolved from bacteria. Some scientists think we're the ancestors of modern animals and plants. I don't know about that. But I do know that all the different kinds of life-forms make for a great, big, interesting world—even for a slime mold.

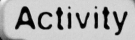

Activity

CLASS GROUPS You've seen that molds can be classified as members of the fungi kingdom, but yellow slime mold is a member of the protist kingdom. Go to the library or Internet to research a mold that's a member of the fungi kingdom, such as mildew. Create a Venn diagram showing the similarities and differences between these two mold species. Focus on how they eat and reproduce, as well as how their cells are structured.

ALMANAC
It's All Relative

Find out how protists and fungi compare to other species on Earth.

Animals may outnumber protists and fungi, but scientists think there are a lot more protists and fungi species waiting to be classified out there. Here's how the kingdoms break down.

NOTE: *Other includes viruses, which some scientists classify as living things.

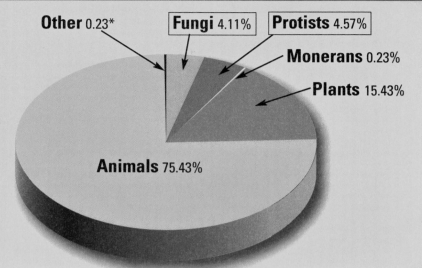

Other 0.23* | Fungi 4.11% | Protists 4.57%
Monerans 0.23%
Plants 15.43%
Animals 75.43%

Under the Rainbow

Fungi aren't just green and brown. Many species come in a variety of colors and textures, though we rarely get to see them. Most grow deep in the forest and other out-of-the-way places.

Scarlet waxy-cap mushroom thrives on soil in woods with evergreens and among deciduous trees, or those that shed their leaves in the fall.

Orange sulfur shelf sprouts from tree trunks and old logs in late summer.

Scrambled egg slime, sometimes known as "the blob," usually grows in wet, grassy areas, but it can spread to other places.

False turkey tail mushroom often grows on decaying trees in the forest. It looks like the turkey tail mushroom, (opposite), but it does not have pores.

Who Moved My Protist?

Some protists move a lot like animals. Study these analogies and see for yourself.

PROTISTS	ANIMALS
A flagellate pulls itself through the water using flagellum, or whiplike extension.	A tadpole propels itself through the water using its tail.
Microscopic ciliates swim using thousands of tiny cilia beating in unison.	The giant Tanzanian millipede can grow up to 2 feet long; it moves using its 200 legs.

This mushroom is named turkey tail because its shape resembles a turkey's tail. Very common, it has bands of color and pores on the underside of its fruiting body.

A very tough, leathery fungus, the winter polypore grows on fallen trees or logs. It's often hard to detach from trees.

Next of Kin

Protists and fungi come in all shapes and sizes. Here are some eye-popping comparisons in these two kingdoms.

FUNGI: Tiny yeast organisms (about 0.003 inch, or 0.075 mm, in diameter) are smaller than the period at the end of this sentence. Yeasts have big siblings: A giant puffball can grow as big as an adult sheep.

size of an average yeast

PROTISTS: A one-celled amoeba, which you can see only under a microscope, is a protist just like the giant kelp in the Pacific Ocean. It can grow up to 200 feet long. If you put one end of a kelp on the goal line of a football field, it would stretch about 20 yards beyond the 50-yard line.

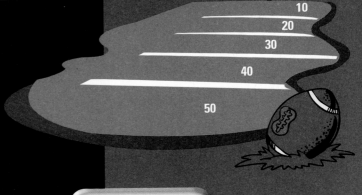

Activity

WATERLOGGED If you've ever cooked a mushroom, you know it shrinks. Fungi are 90 percent water, and when they're exposed to high temperatures, the water evaporates. Go to your local grocery store and buy several kinds of mushrooms. Cook one cup of mushrooms. Measure the cooked mushrooms and estimate how much water evaporated. Why do you suppose there is about 10 percent of the mushroom left?

THE PROTISTS & FUNGI FILES

ALL IN THE FAMILY TREE

How are Indian elephants related to African elephants? Why is an orange different from a banana? Who needs to know, and why?

Indian elephant | African elephant

Scientists do, because they need a way to organize the millions of species of life on Earth. Most scientists accept a system that classifies five kingdoms: animals, plants, monerans (bacteria), protists, and fungi.

Different from all other organisms, bacteria don't have a nucleus, so scientists created two "superkingdoms" based on cell structure: eukaryotes (yoo CARE ee oats)—those with a nucleus—and prokaryotes (pro CARE ee oats)—those without a nucleus. Animals, plants, fungi, and protists are grouped into the eukaryotes because their cells have a nucleus, the central zone containing the cell's genes. Prokaryotes have very simple cells with no nucleus.

Science is always searching for a better classification system. The current five-kingdom system may not be the best. Or is it? Microbiologist Carl Woese wondered about organisms unlike anything scientists had ever seen. His dilemma: How should he classify them?

It's A Small World

Woese understood why most scientists grouped living things into five kingdoms. But he also saw that microbes make up a huge number of organisms living on Earth. (Microbes are unicellular organisms, like bacteria, which are so tiny you need a microscope to see them.) He found that some microbes have unique features. He called these new kinds of microbes archaebacteria (ahr kee bak TIHR ee uh), or archaea (ahr KEE uh) for short. "I was blown out of my mind," says Woese. Then he realized these microbes didn't fit into the five-kingdom tree. He needed a different classification system.

So where did archaea fit? He analyzed their genetic structure and determined that they didn't appear to have evolved from either bacteria or eukaryotes. In 1977 Woese worked out a new tree of life: bacteria, eukarya, and archaea. He combined the animal, plant, protist, and fungi kingdoms into one: eukarya.

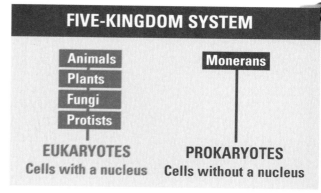

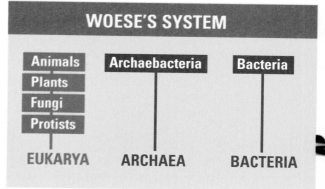

Alien Microbes?

When scientists began studying the areas around hot sulfur springs at Yellowstone National Park, steaming hydrothermal vents on the seafloor, and frozen Antarctic ice, they got a big surprise. Wherever they looked, archaea were thriving! These microbes tell scientists that life may be able to survive in all kinds of harsh conditions—on other planets in our solar system and beyond.

Carl Woese (above) discovered genetic clues in deep-sea vents like those pictured at left.

A Controversial Idea

Many biologists did not accept Woese's new system. They believed archaea was an unusual group that belonged within the kingdom of bacteria.

Woese argued most biologists weren't seeing what he was talking about. "Traditional biologists," Woese says, "focused on animals and plants, the things they could see." He focused on microscopic features that told a different story.

Woese found evidence for his theory a mile and a half deep in the Pacific Ocean. In 1996 scientists unraveled the genetic sequence of an archaea from a deep-sea vent, where the average temperature is very hot—about 176°F (85°C). Most of this organism's genetic structure was unlike anything else on Earth. This organism suggested to Woese that archaea, eukarya, and bacteria shared a common ancestor. And from this universal ancestor came two descendants: Archaea and bacteria were the first two branches, then eukarya split from archaea.

Following this discovery, many scientists accepted Woese's system, although others still disagree, preferring to include archaea as an odd type of bacteria. Scientists may never agree on a single classification system. But as they find new forms of life, there will always be a need to figure out where these life-forms belong.

Activity

A Model Life Look at detailed photographs and diagrams of different cells. Make close observations and give a detailed description of each. Based on your observations, you may want to create your own cell models. Using materials such as Tinker Toys™, pipe cleaners, or other bendable materials, see if you can build a cell with a nucleus. Then try one without a nucleus. Compare the models. How are they alike? How are they different?

THE PROTISTS & FUNGI FILES

PICTURE THIS

Keep on Moving

They might be tiny, but protists need to stay alive just like any other life-form.

Whether you're talking about an elephant or a one-celled Euglena, survival means the same thing: getting enough food to eat, keeping away from predators, reproducing, and adapting to changes in the environment. Take a close look at some protists that have developed adaptations over time to help them survive the game of life. And don't be fooled by the pictures! These protists may all look the same size, but they're not: A paramecium is about 10 times bigger than *Vorticella*.

AMOEBA: Eating on the Go

How do you grab a bite if you don't have a mouth? A one-celled amoeba (a ME ba) uses its membrane both for hunting and capturing bacteria and algae. It moves by pushing out its cell membrane to form fingerlike extensions called pseudopods, or "false feet." Then the jelly-like contents of the amoeba round out the bulge, pulling the rest of the body forward. The amoeba wraps its "feet" around its prey and engulfs the food within its membrane (right). Once inside, the food gets trapped in the amoeba's vacuole (VACK yoo ohl), a cell structure for storing and digesting food.

PARAMECIUM: Making a Fast Getaway

Faster than any other protist, the slipper-shaped paramecium can swim more than two millimeters a second. That means it can travel one centimeter in five seconds—not bad for an organism as small as the dot on an "i." A paramecium uses its speed to outswim predators and sneak up on its prey. It moves by beating thousands of cilia, or hairlike objects, all over its body (left). Together the cilia move the paramecium forward or backward, and they also sweep food into a groove along the organism's body.

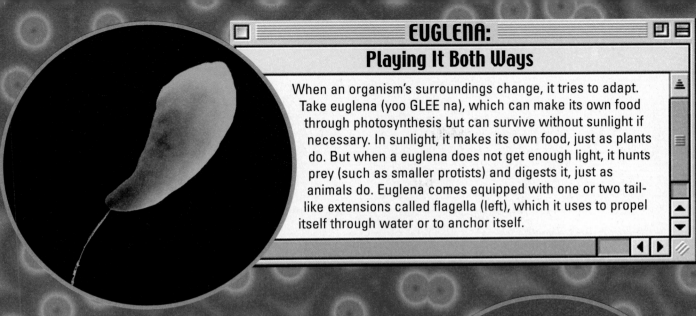

EUGLENA:
Playing It Both Ways

When an organism's surroundings change, it tries to adapt. Take euglena (yoo GLEE na), which can make its own food through photosynthesis but can survive without sunlight if necessary. In sunlight, it makes its own food, just as plants do. But when a euglena does not get enough light, it hunts prey (such as smaller protists) and digests it, just as animals do. Euglena comes equipped with one or two tail-like extensions called flagella (left), which it uses to propel itself through water or to anchor itself.

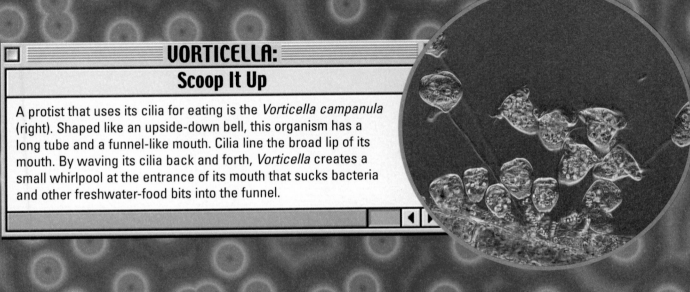

VORTICELLA:
Scoop It Up

A protist that uses its cilia for eating is the *Vorticella campanula* (right). Shaped like an upside-down bell, this organism has a long tube and a funnel-like mouth. Cilia line the broad lip of its mouth. By waving its cilia back and forth, *Vorticella* creates a small whirlpool at the entrance of its mouth that sucks bacteria and other freshwater-food bits into the funnel.

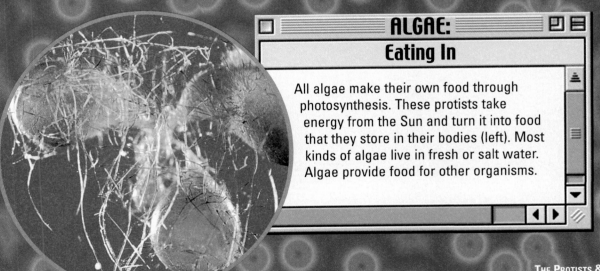

ALGAE:
Eating In

All algae make their own food through photosynthesis. These protists take energy from the Sun and turn it into food that they store in their bodies (left). Most kinds of algae live in fresh or salt water. Algae provide food for other organisms.

THE PROTISTS & FUNGI FILES 13

TIMELINE: HAVING A BREAKDOWN

The life and times of fungi in a compost heap

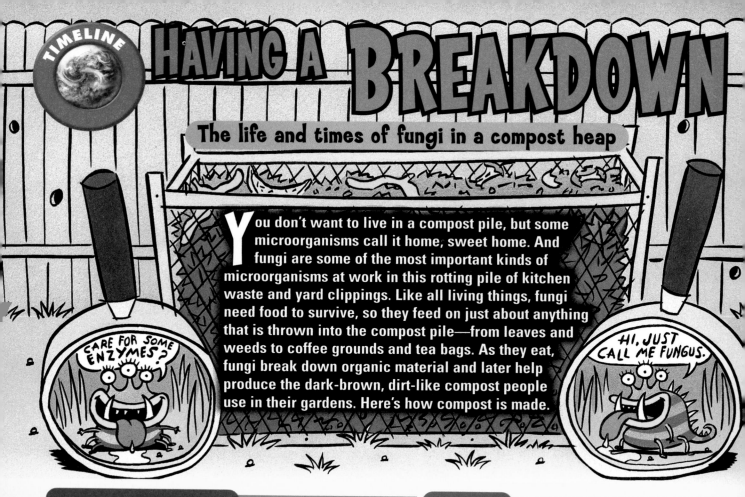

You don't want to live in a compost pile, but some microorganisms call it home, sweet home. And fungi are some of the most important kinds of microorganisms at work in this rotting pile of kitchen waste and yard clippings. Like all living things, fungi need food to survive, so they feed on just about anything that is thrown into the compost pile—from leaves and weeds to coffee grounds and tea bags. As they eat, fungi break down organic material and later help produce the dark-brown, dirt-like compost people use in their gardens. Here's how compost is made.

IN THE BEGINNING: Trash

People throw kitchen and yard waste into the pile. At first the pile is a cool temperature, so medium-heat fungi, which function best in temperatures from 50 to 100°F, go to work. Fungi and bacteria break down the matter into sugars and starches they can eat. Fungi produce enzymes to break down their food. Then they soak it up, the way a sponge soaks up water.

WEEK 1: Reach Out

Fungi have several hyphae (HI fee), or small, thin filaments. To get as much food as possible, hyphae cells divide, making the fungus longer, and branch into many new hyphae. They produce enzymes that break through the cell walls of the organic matter. The fungi absorb the nutrients they need.

WEEK 2: THE COMPOST HEATS UP

With so many fungi and bacteria eating and reproducing at the same time, the compost really heats up. Now high-heat fungi take over. They can survive in temperatures of 100 to 150°F. These fungi eat the stuff the medium-heat fungi and bacteria couldn't absorb—materials like lignen, which gives wood its strength, and cellulose, the material in plants' cell walls.

WEEK 3: THE CYCLE CONTINUES

Fungi and bacteria give off less heat and the compost pile's temperature decreases. The medium-heat fungi finish off the materials that the high-heat fungi can't break down. This is the nutrient-rich compost that people use in their gardens to help plants grow. Later they'll throw more plant matter into a new compost pile. More fungi will grow, and the cycle will continue. It's a good thing we have fungi around to decompose dead matter. One of the by-products of decomposition is carbon dioxide, which helps plants survive. Animals and humans rely on plants to produce oxygen and food. Compost shows us how even the smallest microorganisms like fungi are essential to life on Earth.

Activity

CREATING COMPOST Compost bins are a great way to recycle organic materials. Write a proposal for a compost bin for your school. Before you pitch your plan, consider the following: What is the scientific basis for composting? How do you build a three-chambered compost bin? Where can you obtain earthworms or red worms for the bin? What does your school throw away that can be used for composting? What is the estimated cost for a bin?

Disaster in the Wind

Ireland, Fall, 1845

Some fungi can cause trouble, especially those that kill crops. In fact, a fungus that kills the potato plant caused a terrible famine. The potato was the main source of food for most people living in Ireland in the 19th century. When the fungus destroyed potato crops from 1845–50, nearly one million people died; another million left the country for England or North America.

This fungus feeds off the leaves and stem of the potato plant and also rots the root, which is the part people eat. The fungus reproduces by releasing spores, which travel on the wind and water, so the blight, or disease, spreads quickly from field to field. The fungus first grew in the United States in 1843 and then spread across the Atlantic Ocean two years later, when people brought back infected seeds and planted them in Ireland.

At the time, no one understood that a fungus was to blame. People believed the warm, damp weather was rotting the potatoes. Farmers continued to plant with infected seeds, so the next year's crop failed, too, as well as the crop after that. Here are eyewitness accounts from some who lived through the disaster.

A Fearful Stench

W. Steuart Trench, an Irish farmer in Queen's County, Ireland, heard rumors that the potato blight had spread to his area in August 1846.

> *I rode up as usual to my mountain property, and my feelings may be imagined when before I saw the crop, I smelt the fearful stench, now so well known and recognized as the death-sign of each field of potatoes. . . . [T]he stalks soon withered, the leaves decayed, the disease extended to the tubers [potatoes] and the stench from the rotting of such an immense amount of rich vegetable matter became almost intolerable.*

Starvation and Disease

Although other food goods—meat, dairy products, and wheat—were available at the market, poor Irish farmers had no money to buy them. People already weakened by hunger fell ill from cholera, typhoid fever, and tuberculosis. Diseases swept the countryside. As the rural poor flooded into cities to get government aid, these diseases also killed doctors, relief workers, and others.

The Irish government set up soup kitchens and public work programs, but these efforts were not enough. In 1847 Irish journalist and politician John Mitchel pressured the government to request outside help.

A 19th-century engraving shows a workhouse in Ireland where people came for food.

We know the whole story—the father was on 'public work,' and earned the sixth part of what would have maintained his family, which was not always paid to him, but still it kept them half alive for three months, and so instead of dying in December they died in March. And the agonies of those three months who can tell?—the poor wife wasting and weeping over her stricken children; the heavy-laden weary man, with black night thickening around him—thickening within him— feeling his own arm shrink and his step totter with the cruel hunger that gnaws away his life.

A Soup Kitchen

The United States and other nations responded with donations of food and money. A letter from Elihu Barrit, an American relief worker in Ireland, appeared in the *New York Evening Post* on March 31, 1847. He describes his visit to a soup kitchen.

As soon as we opened the door, a crowd of haggard creatures pressed upon us, and with agonizing prayers for bread, followed us to the soup kitchen. One poor woman had five children sick with the famine fever in her hovel, and she raised an exceedingly bitter cry for help. A man with swollen feet pressed closely upon us, and begged for bread most piteously. He had pawned his shoes for food, which he had already consumed.

A Letter Home

Catherine Hennegan moved to Canada in 1847 to start a new life. In this letter to her parents on February 15, 1848, she is desperate for information about those she left behind.

Dear Father & Mother. I take the present opportunity of letteng you know that I am in good health [,] hoping this will find you and all friends the same. I wrote you shortly after I came here but receivd no answer which make me very uneasy untill I hear from you and how you are and all friends.

By 1850 the situation improved as Irish farmers burned infected fields, switched to different crops, and planted varieties of potatoes that proved resistant to the fungus. An effective pesticide was developed more than 30 years later. But even today this fungus remains a threat to potato crops around the world.

Activity

Fighting Fungus Go to the library or Internet to research the potato fungus *Phytophthora infestans*. Why does it attack potato crops? Is it responsible for other famines? Why is it a difficult fungus to eliminate once it has struck a potato crop?

VIRTUAL VOYAGE

A Paramecium's Paradise

Think a pond is small? In human terms, it is. But to a one-celled protist called a paramecium a pond is an entire universe. Climb into a virtual reality machine and take a trip through the universe, er, pond . . .

Time to start swimming. But in this pond universe you are only one cell. How can a single cell swim with no arms or legs? Feel those thousands of tiny hairs that cover the outside of your cell. They're called cilia, and they move. Swish them quickly back and forth to make yourself slide through the water. It's a little like humans doing the wave at a baseball game. Coordinate your cilia so they move in sequence. That's called beating.

As a paramecium [pair uh MEE see um], you can change your shape and size, but only a little bit. As you beat those cilia, expand your single cell, then contract it. It's like paramecium aerobics, only better! Expand, contract, in, out, in. The change in shape works with the cilia to zip you through the water. Experienced paramecia can move at a rate of about 6 inches per minute. That may not seem fast, but remember, you're smaller than the period at the end of this sentence. You should take what you can get. Most kids can run about 7 miles an hour. As a paramecium you would move closer to 35 miles an hour.

This part of the pond is a great environment for you. The water is still, so lots of plants decay here, breaking down in the water and attracting bacteria, which are a protist's main diet. Now you swim up toward the surface of the pond in search of some. They're all over the place. Up, up, up you go!

Oops. An obstruction is in your way. A stick, maybe, or a lily pad stalk, or even a fishing line. No matter. You throw your cilia into reverse gear, then you push them back and forth. You keep it up, turning ever so slightly until the path is clear.

Lunch 'n' Munch

You know how to swim—you have to, so you can find food. Good thing. You're really getting hungry by now. At last! Here come a few bacteria. You open your gullet, or mouth—it's

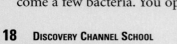

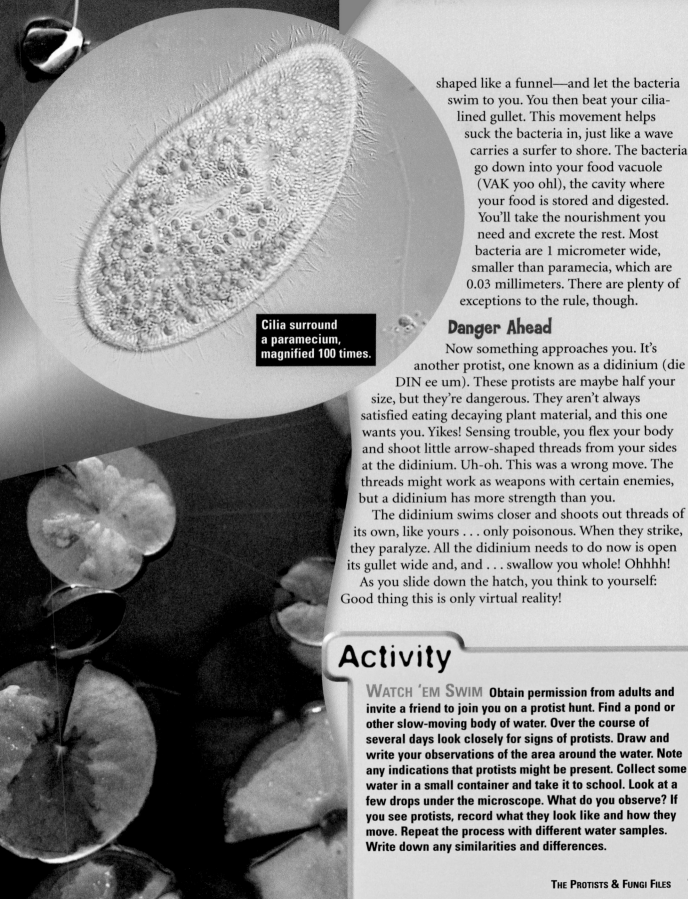

Cilia surround a paramecium, magnified 100 times.

shaped like a funnel—and let the bacteria swim to you. You then beat your cilia-lined gullet. This movement helps suck the bacteria in, just like a wave carries a surfer to shore. The bacteria go down into your food vacuole (VAK yoo ohl), the cavity where your food is stored and digested. You'll take the nourishment you need and excrete the rest. Most bacteria are 1 micrometer wide, smaller than paramecia, which are 0.03 millimeters. There are plenty of exceptions to the rule, though.

Danger Ahead

Now something approaches you. It's another protist, one known as a didinium (die DIN ee um). These protists are maybe half your size, but they're dangerous. They aren't always satisfied eating decaying plant material, and this one wants you. Yikes! Sensing trouble, you flex your body and shoot little arrow-shaped threads from your sides at the didinium. Uh-oh. This was a wrong move. The threads might work as weapons with certain enemies, but a didinium has more strength than you.

The didinium swims closer and shoots out threads of its own, like yours . . . only poisonous. When they strike, they paralyze. All the didinium needs to do now is open its gullet wide and, and . . . swallow you whole! Ohhhh!

As you slide down the hatch, you think to yourself: Good thing this is only virtual reality!

Activity

WATCH 'EM SWIM Obtain permission from adults and invite a friend to join you on a protist hunt. Find a pond or other slow-moving body of water. Over the course of several days look closely for signs of protists. Draw and write your observations of the area around the water. Note any indications that protists might be present. Collect some water in a small container and take it to school. Look at a few drops under the microscope. What do you observe? If you see protists, record what they look like and how they move. Repeat the process with different water samples. Write down any similarities and differences.

SCRAPBOOK: GOOD FOR US

When you hear the words "algae" and "fungi," you might think of slimy seaweed and fuzzy green molds—not exactly what you want for lunch. Yet many kinds of algae and fungi are good enough to eat. And they're good for you, too.

Sea Veggies

About 10 percent of the Japanese diet is seaweed. No wonder—certain kinds of seaweed are rich in vitamins and minerals. Edible seaweed is harvested off the coast of Japan and China, but it's eaten all over the world.

Seaweed	How We Eat It
Nori (NO ri)	Eaten raw; added to sauces and soup; nori leaves are tightly wrapped around pieces of fish and vegetables with rice to make sushi rolls.
Kombu (KOME boo)	Used as a flavoring in fish, meat dishes, and soups; eaten as a vegetable with rice; powdered kombu adds flavor to sauces or soups.
Wakame (wha KAH may)	Used as an additive to soup; coated in sugar and canned as a sweet; eaten with rice.

Gummy Stuff

There's seaweed in your chocolate milk. It's also in your salad dressing. Read the ingredients and look for the word "emulsifier" (ee MUHL si fye er). This gummy substance is made of algin, found in the cell walls of kelp, a brown seaweed. Used in many products, algin keeps things together—from the oil and water in salad dressing to the chocolate and milk in chocolate milk.

Agar, taken from red seaweed, flavors canned meat, fish, and poultry. It is also a thickener in ice cream, pastries, desserts, and salad dressing.

CATCH OF THE DAY

These seaweeds are named for their looks—and are edible, too:

BLADDER WRACK
MERMAID'S FISHING LINE
GRAPESTONE
FAIRIE'S BUTTER
SEA OTTER'S CABBAGE
SEA LETTUCE
SPONGE TANGLE
COW HAIR
SEA NOODLE
STAR JELLY

DISCOVERY CHANNEL SCHOOL

San Francisco, California, 1849

Yeast is the one-celled fungus power behind breadmaking. Mixed into bread dough, yeast uses the dough's sugar to produce carbon dioxide gas in the process of fermentation. The carbon dioxide trapped inside the dough makes it rise.

During the California Gold Rush of 1849, pioneers traveling west often carried a sourdough starter—a mixture of flour and water—in small leather bags around their necks. They used the starter to make bread and earned the nickname "sourdoughs" for their strong-smelling yeast bags.

Fungus Feasts

Most cheeses aren't ready to eat until they've had enough time to ripen, or age. A cheese's special flavor develops from mold growing inside it. Mold also forms the cheese rind, or protective outer layer.

Cheese is made from the solids after the liquid is removed from milk. Added to the solids, mold gives the cheese its flavor. Some cheese molds are easy to see, like blue cheese. Check these out:

- Blue cheese gets its flavor from a mold similar to the one used to make penicillin. Spores of this mold are mixed in with the curds, or clumps of thickened milk. As the cheese ripens for three to six months, the bluish mold grows in cracks and spaces in the cheese. The cheese maker controls the growth of the mold by controlling the temperature of the cheese's storage place.

- Camembert (CAM em bear) cheese has a thick white rind made of mold that protects the creamy cheese inside. Both inside and outside parts are edible.

Back to Basics

In the novel *My Side of the Mountain*, by Jean Craighead George, the main character, Sam, runs away from home. He lives in a hollowed-out tree and harvests his food from the forest.

With pockets and good tough pants I was willing to pack home many more new foods to try. Daisies, the bark of a poplar tree that I saw a squirrel eating, and puffballs. They are mushrooms, the only ones I felt were safe to eat, and even at that, I kept waiting to die the first night I ate them. I didn't so I enjoyed them from that night on. They are wonderful. Mushrooms are dangerous and I would not suggest that one eat them from the forest. The mushroom expert at the Botanical Gardens told me that. He said even he didn't eat the wild ones.

Activity

FIND THE HIDDEN INGREDIENTS

Now that you know protists and fungi are everywhere, take a thorough inventory of your home. Start in the kitchen and make a list of objects that might have some kind of protist or fungi among its ingredients. Check ingredients labels for emulsifiers, seaweed, yeast, and dried mushrooms. Next check ingredients listed on items in the medicine cabinet. If you live in a house with a yard, check outside for any wild mushrooms. Use your list to make a chart or Venn diagram, grouping items by whether they contain protists, fungi, or both.

AMAZING BUT TRUE

TO EAT or

DEADLY 'SHROOMS

If an apple is the fruit of an apple tree, then what's a mushroom? It's the fruit of a fungus! It's not really a fruit because it doesn't have seeds, but it does have thousands of single-celled spores. Lots of so-called fungus fruiting bodies are edible, but many more are not. Fungi can be prized and quite tasty, but others are downright poisonous. Mycologists, or people who study fungi, can identify wild mushrooms for safety before people eat them.

Every year in the United States, hundreds of people get mushroom poisoning because they eat wild mushrooms without consulting experts. On average, about a dozen people suffer severe poisoning, and a few cases are fatal.

In 1767, the French composer Johann Schobert and his family picked wild mushrooms in their village near Paris. The chef at a local restaurant refused to cook the mushrooms because he thought they were poisonous. So Schobert cooked them himself at home. Later Johann, his wife, and most of his children died because they had eaten the deathcap, a type of *Amanita* mushroom. Another deadly *Amanita* is the destroying angel. Those who eat this tasty mushroom don't get sick for a few hours. Then they suffer from kidney or liver damage, which often causes death.

But some kinds of *Amanita* mushrooms are not poisonous. The rulers of ancient Rome enjoyed Caesar's mushroom. Many experts recommend not eating any of the *Amanita* mushrooms, though, because they all look alike. Here a case of mistaken identity can be fatal.

Can you tell which mushrooms are poisonous?

A Deathcap

B Caesar's mushroom

C Destroying angel

Answer: A and C are poisonous.

Not to Eat

Fit for a Pharaoh?

In ancient Egypt some fungi were so prized that only royalty were allowed to eat them. Today many varieties of edible fungi are rare. Most are collected in nature and sell for between $20 to $130 a pound.

Truffles, one famous prized fungus variety, are rare and difficult to find. The truffle has a unique partnership with pines, firs, and oaks. It grows on the tree roots, absorbing nutrients from the tree while helping the tree absorb minerals in the soil. Since truffles grow underground and have a very distinct smell, specially trained pigs and dogs—with their highly developed sense of smell—help find them. In Italy truffles are so valuable that collectors must carry a license.

Some hunt morels (more ELZ), too. Every spring people search woodlands for edible morels; they grow for just a few weeks. Unlike truffles, morels grow above ground and are easier to find. While most are edible, an entire class of "false" morels are poisonous, so people must know how to tell one variety from another.

Specially trained pigs help find truffles (below).

The morel *Morchella deliciosa*

Activity

PLAYING THE ODDS Sometimes it pays to use a little math! There are about 1,000 known mushroom species on Earth. About 70–80 are poisonous. Given this, what is the chance that you'll find a mushroom that is not safe to eat? Calculate your answer as a percentage.

MAP: The Protozoan Zone

Some protists are tiny, but they sure do get around. Consider the protozoa, one-celled protists that thrive in seawater, freshwater, soil, and other moist areas—even in cells and fluids of other organisms. Many protozoa are parasites, meaning they live inside hosts, or other organisms. Some protozoa are harmless, but others are deadly and cause disease in every corner of the world by spreading through insects, animal waste, unclean water, and contaminated food.

MAP KEY
- Malaria
- Giardiasis
- Crypto
- Crypto Parvum

◆ Crypto Parvum
Sometimes a protozoan can strike drinking water. In 1993 a species of *Cryptosporidium* (krip tow spo RIH dee um) made its way into the water supply of Milwaukee, Wisconsin. About 403,000 people fell ill with *Cryptosporidium parvus*, or crypto parvum. High temperatures kill this parasite, so people often boil drinking water during such an outbreak.

Malaria
Malaria is the world's worst tropical disease. Blame it on four species of *Plasmodium*, a protozoan that grows inside the *Anopheles* (an OFF el eez) mosquito. When a female mosquito bites someone, the protozoan moves into the person's bloodstream, reproduces, and causes illness. Malaria is a big public health problem in developing countries, where mosquitoes breed in ditches, swampy areas, and standing water.

Giardiasis
The next time you think of drinking from a babbling brook or a cool stream, hold up! Even water that looks crystal clear can contain parasites. The parasite giardia lives in many streams and lakes. It causes giardiasis (jee ahr DEYE uh sihs), a digestive disorder most commonly caused by a protozoan. To stay healthy, campers and hikers boil drinking water or purify it with iodine.

KISS OF DEATH

Chagas
Watch out for the kissing bug: It bites the lips, eyelids, or ears of a person sleeping. The insect carries a trypanosome protozoan and spreads the disease called chagas (SHAH gus) to humans through its waste. Inside a human host, the protozoan damages the heart. Chagas is the main cause of heart failure in South and Central America. As the rain forest is cut down, kissing bugs are losing their habitat. Many now live in cracked walls and thatched roofs in houses bordering the rain forest.

Crypto Some protozoa spread to people who live and work closely with animals. Take *Cryptosporidium*, which breeds in cows, dogs, or cats. People can come down with cryptosporidiosis—crypto, for short—if they are exposed to animal waste. The protozoan causes diarrhea, fever, and dehydration. It is particularly bad for people such as the elderly and AIDS patients, who have weak immune systems.

African Sleeping Sickness Like the kissing bug, the African tsetse (TEET see) fly carries a trypanosome protozoan. It enters the bloodstream through a bite; there it multiplies and damages the nervous system. It's called sleeping sickness because it makes people extremely tired; it also causes sore muscles and severe headaches. Although treatable, sleeping sickness kills many people in areas that lack well-equipped health clinics or hospitals.

Activity

DISEASE AND NUMBERS Research one of the diseases shown on the map. Answer the following questions:
1. In what 10 countries is the disease most prevalent?
2. How many cases have been reported in each country in the past five years?
3. When was the last major outbreak? Are reported cases decreasing or increasing?
4. What percent of the population is affected each year? What is the death rate from this disease?

The Miracle of Mold

London, 1928

Sometimes it pays to be messy. Scottish scientist Alexander Fleming left out old lab dishes for weeks. One day he came across a dish covered with bacteria and a blue mold. Some scientists would have cleaned it or thrown it away, but not Fleming. This was a mold that would change history.

Fighting Bacteria

Alexander Fleming began his career as a research assistant to Sir Almroth Wright, a doctor who developed vaccines. Fleming studied the body's reaction to bacteria and how doctors treated infected wounds. In the 1920s few treatments existed to stop bacterial infections.

During World War I (1914–18) Fleming worked in a hospital lab on a battlefield in France. There he saw many soldiers dying of infected wounds. Doctors treated soldiers with antiseptics—substances that prevent the growth of microorganisms, like bacteria. But antiseptics killed more healthy cells than bacteria. Often an injured soldier got sick even from minor wounds, and the antiseptics were useless once a wound was infected.

After the war Fleming searched for a substance that would kill bacteria and fight infection without damaging healthy cells. He believed he could find a substance to fight infection safely. It was just a matter of finding and testing it.

> "Nature makes penicillin. I just found it."
> —Alexander Fleming
> (after his breakthrough discovery)

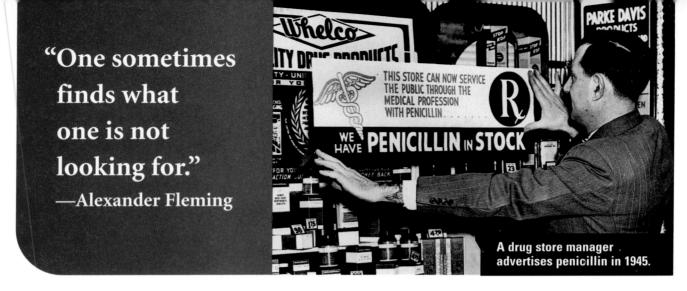

"One sometimes finds what one is not looking for."
—Alexander Fleming

A drug store manager advertises penicillin in 1945.

The Perfect Mold

A decade after the war, in 1928, Fleming's hunt came to an accidental end—right in his own lab. In a dirty petri dish, a small lab dish scientists use for studying microorganisms, a layer of bacteria grew. Alongside it was a tiny blue mold. Fleming observed that no bacteria grew on the mold: It appeared that the mold stopped the bacteria from growing. Perhaps this mold—*Penicillium*—could fight all bacteria.

Fleming went to work. He discovered that the mold releases a chemical that is poisonous to bacteria, but is not harmful to the body. He named it penicillin. It could treat infected wounds, and it had the potential to fight many other bacterial infections. But Fleming did not have the training to make the mold into a medicine.

Breakthrough!

At the beginning of World War II, biochemists Sir Howard Florey and Ernst Boris Chain at Oxford University supplied the missing piece. They set to work purifying the mold Fleming had first observed in his lab. By the mid-1940s they succeeded in creating the first antibiotic, or antibacterial drug, in the world.

In 1945 Fleming shared the Nobel Prize with Chain and Florey for developing penicillin. The drug saved many lives during World War II and has fought pneumonia, scarlet fever, diphtheria, and other diseases. Today penicillin helps control many harmful diseases. Antibiotics have made most surgery safer for patients and have helped prevent the spread of many fungus infections.

Thanks to Fleming's careful observation, penicillin is the most widely used antibiotic in the world.

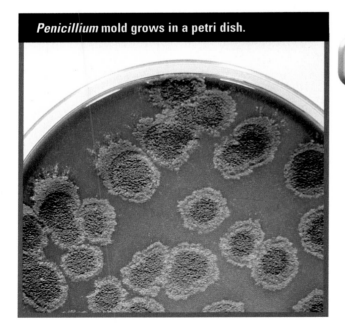

Penicillium mold grows in a petri dish.

Activity

LIFE SAVER Even though penicillin wasn't developed until the last year of World War II, it made a great difference in the number of lives that were saved. You can appreciate Fleming's contribution especially when you look at statistics from World War I. Choose a disease, such as scarlet fever, diphtheria, or pneumonia, and go to the Internet or the library to learn how many people died from the disease between 1914–18, and how many died after penicillin's discovery. If possible, organize the statistics by continent. Put your information in a table, comparing statistics in each time period and in each part of the world. Write three or four concluding statements based on your data.

The Mystery of the Failing FISHING HOLE

"I can't wait to see some seals and sea lions," shouted Marnie Fisher above the engine roar of her grandfather's boat. "I don't see them as much back home in California any more. I guess they're finding new homes or something." Marnie peered through her binoculars along the rocky shores of the Aleutian Islands. "If I were a seal, I'd definitely move here."

Grampa Fisher slowed the boat down and checked his readings. "We're gettin' close to my top-secret favorite fishing spot. I expect we'll be seeing seals and sea lions any moment now. Look over there—" he pointed to a golden brown lump on a distant rock.

"I see him!" Marnie said, a huge grin growing on her face as she zeroed in on the lone sea lion.

Grampa cut the engine. "This is the place! My special fishing spot that's been good to me for over 40 years. I must admit I didn't recognize it right away. There used to be a lot of kelp and algae here for the fish. It's changed so much in the past several years." He stepped onto the bow and took in the scenery. "I've always looked for the seals to tell me when I'm getting close to my spot. Now there are hardly any. There used to be so many more."

"I wonder where they've gone," said Marnie, her smile fading.

"Let's ask the fish," joked Grampa, handing her a fishing pole. "Even if we don't see seals, I expect we'll be catchin' an amazing number of fish today!"

Marnie and her grandfather spent the rest of the afternoon casting lines and pulling in a lot of nothing. Marnie experimented with letting her line out at different depths, but she didn't get a nibble all day.

CLUES

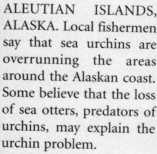

ALEUTIAN ISLANDS, ALASKA. The U.S. Fish and Wildlife Service reports that the sea otter population has declined by 95% since the 1980s. Studies show that pollution is one contributor to sea otter deaths.

ALEUTIAN ISLANDS, ALASKA. Local fishermen say that sea urchins are overrunning the areas around the Alaskan coast. Some believe that the loss of sea otters, predators of urchins, may explain the urchin problem.

SEA OTTER FACTS
HABITAT: Pacific coastal areas from Alaska to Baja, California
DIET: fish, sea urchins, and small mammals

SEA URCHIN FACTS
HABITAT: tide pools and kelp forests, outer Pacific coastal areas from Alaska to Baja, California
DIET: mainly algae; seaweed such as kelp; and animals

Grampa looked down at the water and frowned. "Seems like every year I'm seeing more and more of those sea urchins. If we could have a fish for every urchin I see, we'd have the catch of the season!"

"Don't sea otters eat sea urchins?" asked Marnie.

"It's one of their favorite foods," nodded Grampa. "They eat 'em all day long. In fact, we should head back out so that we don't drift in too close to the kelp beds that surround this area. Keep a keen lookout with the binoculars and let me know if you spot any sea otters. They're always a sure sign that kelp beds are nearby."

"Nothing, Grampa—the coast is clear," shouted Marnie. But that seemed to be the problem: The coast was clear. No seals, no fish . . . and no kelp beds or otters, either.

That evening they shared their stories with Grandma Fisher. "Well, I can tell you why at least one otter is missing," she said, pointing to an article in the newspaper. "A killer whale's been spotted attacking sea otters again."

"It's happened before?" asked Marnie.

"See for yourself," said Grandma, handing her the article.

Marnie looked it over. "Hmm, it says here that the decline in seal and sea lion populations has caused killer whales to prey on sea otters for food. Something just doesn't seem right about this story. Why are the seals disappearing?"

"You'd think with fewer seals, there'd be more fish for you and Grampa," commented her grandmother.

"I'm going to do some detective work and see if I can figure out what is happening," announced Marnie.

By the end of the week, Marnie had a whole stack of news stories and library books. "Grampa! I think I've figured out why there aren't any more fish in your secret fishing spot!"

Grampa Fisher looked over Marnie's notes and newspaper clippings. "You've certainly found some interesting facts, but I don't see how any of them are connected. And what do these things have to do with my fishing spot?"

What did Marnie tell her grandfather? Look at her research in the clues below to solve the mystery.

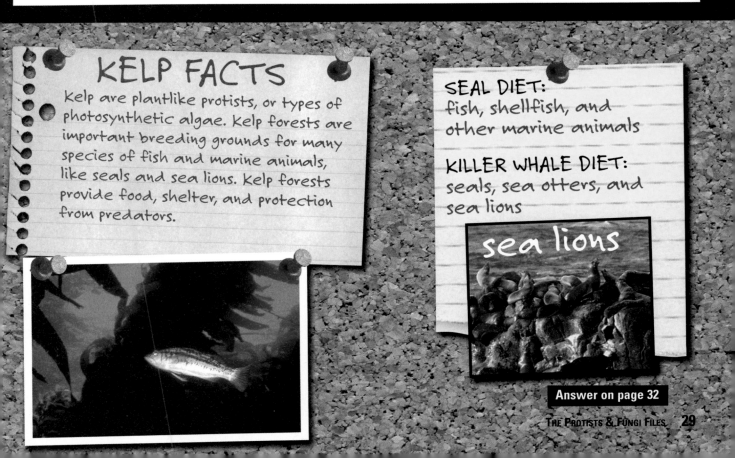

KELP FACTS

Kelp are plantlike protists, or types of photosynthetic algae. Kelp forests are important breeding grounds for many species of fish and marine animals, like seals and sea lions. Kelp forests provide food, shelter, and protection from predators.

SEAL DIET:
fish, shellfish, and other marine animals

KILLER WHALE DIET:
seals, sea otters, and sea lions

sea lions

Answer on page 32

Slimy Stuff

Q: Why did the mushroom go to the party?

A: Because he's a fungi (FUN guy).
(Note: Fungi is usually pronounced FUN-jeye)

Q: Why didn't the fungus go to the party?

A: Because there wasn't mushroom.

A Humongous Fungus

Researchers found the largest fungus in the world, called *Armillaria ostoyae*, under the soil in eastern Oregon in August 2000.

Size: covers 2,400 acres, or about 1,665 football fields

Age: possibly 2,400 years old

How it gets around: slowly grows by extending the tips of its filaments, creeping from tree to tree along roots.

Slime Maze

Slime molds are protists—single-celled masses that creep along and feed on bacteria all day. But can slime molds think? Researchers in Japan believe so. To test their theory, they put a slime mold in a maze. Its challenge: to find some oat flakes. The researchers set up the maze with the food at different exit points. As the slime mold moved through the maze, it changed its shape, shrinking when it reached a dead end. Once it found the shortest route between the two blocks, the mold quickly moved toward the food. So slime molds know this, too: The shortest distance between two points is a line.

Invasion of the Red Blob!

A man discovered a pulsating red blob in his backyard in Dallas, Texas, in 1973. News reporters called it a new form of life, but biologists identified it as a slime mold. At 3 feet wide and 30 feet long, it was one of the largest slime molds ever found.

THE BIG CHEESE

Even queens like mold. A kind of fungus, mold is used to make cheese. In 1837 Queen Victoria of England was presented with a giant cheese. It stood 20 inches (50.8 cm) high, measured more than 9 feet (2.74 m) in diameter, weighed 1,100 pounds (499 kg), and required one full day's yield of milk from 78 cows. Some fungi really get the royal treatment!

YUCK— IT'S BETWEEN MY TOES!

Next time you're in the locker room, watch out for athlete's foot. Fungus causes this skin disease on the bottom of the foot and between the toes. Shoes create a warm, dark, and humid environment, a perfect place for this fungus to grow.

Places to catch it: swimming pools, showers, and locker rooms

Symptoms: drying skin; blisters; itching and burning

How to avoid it: Wash your feet. Use soap and water every day, and dry your feet completely, especially between your toes; change your shoes and socks to reduce moisture; use foot powder if needed.

I'LL HUFF AND I'LL PUFF . . .

You may have accidentally stepped on a puffball sometime. Puffballs are round fungi bodies that contain spores. When you step on them, spores go "puff" and fly away. Giant puffballs can measure more than 2 feet long and contain 7 trillion spores.

EARTH STARS

In the rain the many fleshy layers of an earth star fungus uncurl and bend outward to reveal its fruiting body. That's when the fungus looks like a star.

FAIRY RINGS

Some mushrooms seem to come out of nowhere. Take the fairy ring, a circle of toadstools, or inedible mushrooms, that can show up on your lawn overnight. Long ago people believed these were magic circles that fairies left behind. Today we know that the rings grow from hyphae (HI fee), the threads that form a fungus body. The hyphae grow outward in all directions to produce a ring.

THE LOVIN' LICHEN

Lichen (LYE kin) is the marriage between an alga and a fungus. These two grow together, often on rocks in a forest, where it's nice and damp, perfect for a fungus. The two usually can't live without each other. The alga makes food for the fungus, and the fungus provides water and minerals as protection for the alga.

Scientists believe that millions of species of fungi and protists may exist. They discover new kinds of protists and fungi every day.

To make sense of the natural world, people need systems of classification that help organize things into groups. Yet not everyone agrees on how to do this. Should you group species into kingdoms based on their physical descriptions or their structure and function? Are there other ways to classify your world?

Some scientists believe there are as few as three kingdoms, while others think there are as many as twelve. Decide for yourself. Form teams. Go to the library and research different classification schemes, or systems, that scientists use today. After you've completed your research find one scheme that the entire group supports. Choose from these systems: 3, 5, 7, 9, or 12 kingdoms.

Debate with classmates why your classification scheme makes sense.

Before you debate consider the following:

1. What is a simple way to describe the organisms in each kingdom?
2. By what characteristics are organisms classified?
3. What makes each kingdom in this system different?
4. How would you compare your system with other classification systems? How is your system more effective than others? How is it less effective than others?
5. Why would school books use this system?
6. Why would scientists use this system?

Ready for the ultimate challenge? Enter this or any other science project in the Discovery Young Scientist Challenge. Visit http://school.discovery.com/sciencefaircentral/dysc/index.html to find out how.

ANSWER Solve-It-Yourself Mystery, pages 28–29

Sea otters were declining and that's why Grampa Fisher's favorite fishing spot had no fish! Without sea otters, the sea urchins had no predator, so urchins multiplied and gobbled up all the kelp and other algae.

Animals are consumers. They get their energy from producers, such as algae and plants. An alga, like kelp, is a kind of protist. It provides food for many animals. When the kelp and other algae disappeared, this caused a real problem for the fish. They lost their habitat and food.

Without kelp—and fish—seals and sea lions moved away. The whales, who usually rely on seal and sea lions for food, were forced to eat the sea otters instead. This fact—and the problem of pollution—explains why there are fewer sea otters.

DISCOVERY CHANNEL SCHOOL